Invisible Journeys
Energy

Caroline Grimshaw

TEXT EDITOR IQBAL HUSSAIN
SCIENCE CONSULTANT JOHN STRINGER

World Book
in association with
WWCN

Invisible Journeys
Energy

CREATIVE AND EDITORIAL DIRECTOR
CONCEPT/FORMAT/DESIGN/TEXT
CAROLINE GRIMSHAW

TEXT EDITOR **IQBAL HUSSAIN**
SCIENCE CONSULTANT **JOHN STRINGER**
ILLUSTRATIONS
NICK DUFFY ✻ **SPIKE GERRELL**
CAROLINE GRIMSHAW

THANKS TO
TIM SANPHER COMPUTER IMAGERY
LAURA CARTWRIGHT PICTURE RESEARCH
MELISSA TUCKER U.S. EDITOR, WORLD BOOK PUBLISHING

TITLES IN THIS SERIES →
✻ SUN
✻ COMMUNICATION
✻ SOUND
✻ ENERGY

PUBLISHED IN THE UNITED STATES BY
WORLD BOOK, INC., 525 W. MONROE, CHICAGO, IL 60661
IN ASSOCIATION WITH TWO-CAN PUBLISHING LTD.

© CAROLINE GRIMSHAW/TWO-CAN PUBLISHING LTD., 1998.
ALL RIGHTS RESERVED. NO PART OF THIS PUBLICATION MAY BE REPRODUCED, STORED IN A RETRIEVAL SYSTEM, OR TRANSMITTED IN ANY FORM OR BY ANY MEANS ELECTRONIC, MECHANICAL, PHOTOCOPYING, RECORDING OR OTHERWISE, WITHOUT PRIOR WRITTEN PERMISSION FROM THE PUBLISHER.

FOR INFORMATION ON OTHER WORLD BOOK PRODUCTS,
CALL 1-800-255-1750, EXT. 2238, OR VISIT US AT OUR WEB SITE AT
HTTP://WWW.WORLDBOOK.COM

LIBRARY OF CONGRESS CATALOGING-IN-PUBLICATION DATA
GRIMSHAW, CAROLINE.
  ENERGY / CAROLINE GRIMSHAW; TEXT EDITOR IQBAL HUSSAIN, SCIENCE CONSULTANT JOHN STRINGER; [ILLUSTRATIONS NICK DUFFY, SPIKE GERRELL AND CAROLINE GRIMSHAW].
     P. CM. -- (INVISIBLE JOURNEYS)
  INCLUDES INDEX.
  SUMMARY: QUESTIONS AND ANSWERS AND ACTIVITIES EXPLORE THE DEFINITION, CAUSES, EFFECTS, AND USES OF ENERGY.
  ISBN: 0-7166-3004-4 (HC.) -- ISBN 0-7166-3005-2 (SC.)
  1. FORCE AND ENERGY--MISCELLANEA--JUVENILE LITERATURE. 2. POWER RESOURCES--MISCELLANEA--JUVENILE LITERATURE. 3. POWER (MECHANICS)--MISCELLANEA--JUVENILE LITERATURE. [1. FORCE AND ENERGY--MISCELLANEA. 2. POWER RESOURCES--MISCELLANEA. 3. POWER (MECHANICS)--MISCELLANEA. 4. QUESTIONS AND ANSWERS.]
  1. DUFFY, NICK, ILL. II. GERRELL, SPIKE, ILL. III. TITLE. IV. SERIES.
QC734.G75 1998
333.79--DC21   98-3577

PRINTED IN SPAIN

HARDBACK 1 2 3 4 5 6 7 8 9 10 02 01 00 99 98
PAPERBACK 1 2 3 4 5 6 7 8 9 10 02 01 00 99 98

I am your Route-Scout. I will show you the way. Look out for my two companions on your journey.

Welcome TO Invisible Journeys

THE Highway

Travel along the Highway following the journey of energy from its source to its end – the use of the energy itself.

THE Side Roads

On your journey you will be asked to select your own route. Choose a Side Road and follow the path.

THE Road Stops

The Side Roads lead you to Road Stops, which contain vital information about your trip. These may lead you farther – watch out for Points of Interest, which are bursting with fascinating facts. Visible Proof Spots will test your knowledge with experiments and puzzles. Detours allow you to leap forward to Road Stops farther along the route. They have a symbol that looks like this. ------------------→

Let's examine energy!

Detour

A journey that powers our planet

FOLLOW THE **HIGHWAY** – WE'RE ON OUR WAY!

A journey that makes plants grow, people move, and machines work

Light, heat, sound, electricity – all of these are forms of energy

FOLLOW THE **HIGHWAY**. →

Our journey begins with energy itself – what it is, what forms it takes, and its role in the universe!

Energy gives life to the planet. Let's find out more!

Select
YOUR SIDE ROAD

UNDERSTANDING Energy

1 What is energy?
(HOW DO WE MEASURE IT?) — SIDE ROAD TO ROAD STO[P]

2 In what different forms is energy found?
(CAN ENERGY BE STORED?) — SIDE ROAD TO ROAD STOP 2

3 Where does the earth's energy come from?
(WHERE DOES THE SUN GET ITS ENERGY FROM?) — SIDE ROAD TO ROAD STOP 3

4 What is the big bang – and what does it have to do with energy? — SIDE ROAD TO ROAD STOP 4

5 What is the big crunch? — SIDE ROAD TO ROAD STOP 5

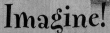

Imagine!
Our universe may have been created by a massive explosion of energy, called the big bang. Boooom!

ROAD STOP 1

What is energy?

Energy is all around us. It is usually invisible, but it makes all things happen. Nothing can live, move, work or change without energy.

UNDERSTANDING ENERGY

On some days, you may feel full of energy. You think that you can make all kinds of things happen – from hopping and skipping to running and jumping. The more energy you have, the more you can do.

WORK, FORCE AND POWER

❋ **WORK** = the capacity of energy to make things happen, or to do work. Work cannot be done without energy. The word "energy" comes from a Greek word meaning activity.

❋ **FORCE** = a push or a pull on an object. Work is a result of a force moving an object.

❋ **POWER** = the rate at which work is done.

ENERGY IS EVERYWHERE

 ❋ A cheetah needs energy to run. It takes its energy from the animals that it eats for food.

 ❋ A windsurfer uses the energy of the wind and the waves to sail across the water.

Visible Proof SPOT

If you hold a ball up in the air, the ball is not doing work, because it is not moving. There is no energy for work to be done. If you throw the ball to a friend, work is done because the ball moves in the direction you throw it. Your muscles supply the energy needed.

Detour
FIND OUT ABOUT THE DIFFERENT FORMS OF ENERGY. LEAP TO **ROAD STOP 2**.

ENERGY MAY BE INVISIBLE, BUT WE CAN STILL MEASURE IT. FOLLOW THE PATH TO THE **POINT OF INTEREST**.

---- TO FIND OUT ABOUT ENERGY AND PEOPLE, FOLLOW THE **HIGHWAY**. ---->

POINT of Interest
Let's take a look at the units we use...

How do we measure energy?

Energy is measured in units called joules (J), after the English scientist James Joule (1818-1889). There are 1000 joules in 1 kilojoule (kJ).

You need to do 1 joule of work to lift an orange into the air by 3.3 ft. (1 m).

A whole orange contains about 60 kcal, or 252 kJ, of energy.

KILOCALORIES

The amount of energy in food is measured in kilocalories (kcal) or kilojoules. One kilocalorie is equal to about 4.2 kilojoules.

Detour
FIND OUT MORE ABOUT FOOD ENERGY. LEAP TO **ROAD STOP 8**.

Visible Proof SPOT

The more active you are, the more energy you need. Examine this chart.

ACTIVITY	ENERGY USED PER HOUR
SLEEPING	220 KJ (50 KCAL)
READING	430 KJ (100 KCAL)
WALKING	900 KJ (210 KCAL)
RUNNING	2,880 KJ (690 KCAL)

---- WHAT IS KINETIC ENERGY? **SIDE ROAD** TO ROAD STOP 2
---- HOW DOES THE SUN MAKE ENERGY? **SIDE ROAD** TO ROAD STOP 3
---- WHAT IS THE BIG BANG? **SIDE ROAD** TO ROAD STOP 4
---- WHAT IS THE BIG CRUNCH? **SIDE ROAD** TO ROAD STOP 5

ROAD STOP 2

In what different forms is energy found?

There are many kinds of energy, including kinetic, electrical, heat, sound, light, and nuclear energy. The different forms of energy make different things happen.

This rower is turning energy stored in his muscles into kinetic energy.

Visible Proof SPOT

If the speed of a moving object doubles, its kinetic energy increases four times. If a toy train moving at 10 ft. (3 m) per second has 6 J of kinetic energy, how much energy will it have if it moves at 20 ft. (6 m) per second?

ANSWER: 24 J.

1 KINETIC ENERGY

The word "kinetic" comes from the Greek word *kine*, meaning movement. All moving objects have kinetic energy. The bigger the object and the faster it moves, the more kinetic energy it has.

2 ELECTRICAL ENERGY

Electrical energy, or electricity, occurs with the flow of tiny particles, called electrons, through substances.

WHAT IS NUCLEAR ENERGY? SIDE ROAD TO ROAD STOP 2

FOLLOW THE HIGHWAY.

ROAD STOP 3

Where does the earth's energy come from?

Most of the earth's energy comes from the sun. Most living things, including people, animals, and plants, depend on this energy. The energy travels about 90 million miles (150 million km) and arrives at the earth as sunlight, giving both light and heat.

Detour

RADIATION CAN BE DANGEROUS. DETOUR TO ROAD STOP 29.

RADIATION ENERGY

Light and heat are examples of radiation energy, often called radiation. Other types of radiation include X rays and radio waves.

※ Less than one-billionth of the sun's energy reaches the earth.
※ The sun transmits more energy to the earth in one hour than all of the world's people use in one year.

FOLLOW THE PATH TO THE POINT OF INTEREST TO FIND OUT ABOUT NUCLEAR FUSION.

SIDE ROAD TO ROAD STOP 2
SIDE ROAD TO ROAD STOP 3
SIDE ROAD TO ROAD STOP 4
SIDE ROAD TO ROAD STOP 5

3 LIGHT, HEAT, AND SOUND

✳ Light, heat, and sound are all types of energy that travel in waves.
✳ Unlike heat and light waves, sound waves cannot travel in a vacuum – a space from which all air or other substance has been removed.
✳ All the different forms of energy can be changed to heat.

4 NUCLEAR ENERGY

A huge amount of energy is stored in the center, or nucleus, of an atom. The energy is released either when the nucleus of an atom splits (fission), or when two or more nuclei join together (fusion).

FOLLOW THE PATH TO THE **POINT OF INTEREST** TO FIND OUT ABOUT TWO MORE TYPES OF ENERGY.

Electrical energy flows through a television set, where it is changed to light, sound, and heat.

Detour
FIND OUT MORE ABOUT HOW ONE FORM OF ENERGY IS TRANSFORMED INTO ANOTHER. LEAP TO **ROAD STOP 15**.

POINT of Interest
Some energy is stored – ready for action!

Can energy be stored?

Yes. Many things have a supply of energy stored within them that can be used at a later time.

POTENTIAL ENERGY

Potential energy is the stored energy that an object has because of its position. A rock perched on a cliff has potential energy. An object, such as a spring, gains potential energy if it is twisted, squeezed, or stretched.

CHEMICAL ENERGY

Chemical energy is a form of potential energy stored in objects such as food, fuel, plants, and batteries. This energy is released by chemical reactions. Chemical changes inside a battery release the stored energy to produce electricity.

----- FOLLOW THE **HIGHWAY** TO FIND OUT HOW SUNLIGHT SUPPLIES THE ENERGY THAT PLANTS NEED TO GROW. ----→

POINT of Interest
Fusing together to make energy.

Where does the sun get its energy from?

The sun releases its energy from powerful nuclear reactions, using a process called nuclear fusion.

Detour
COULD THE SUN'S ENERGY BE A FUEL OF THE FUTURE? LEAP TO **ROAD STOP 30**.

UNDERSTANDING ATOMS

Almost everything is made up of atoms. Each atom is made up of a nucleus surrounded by electrons.
NUCLEUS = MADE UP OF TINY PARTICLES CALLED PROTONS AND NEUTRONS
ELECTRONS

The sun gets energy from nuclear fusion reactions deep in its center. Atoms of a gas called hydrogen collide and combine to form atoms of another gas, called helium. As they join together, or fuse, they release neutrons and a vast amount of energy.

Visible Proof SPOT

The tiny nucleus is a very small part of the atom. If the nucleus was the size of this cube, the atom would be over half a mile (1 km) in diameter!
.4 IN. (1 CM)

HOW NUCLEAR FUSION HAPPENS

Nuclear fusion has to take place in extreme heat and with fast-moving, lightweight nuclei.

[1] Two or more hydrogen nuclei smash together.

[2] They join to form a heavier, helium nucleus.

[3] This releases energy and neutrons.

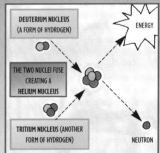

DEUTERIUM NUCLEUS (A FORM OF HYDROGEN)
THE TWO NUCLEI FUSE CREATING A HELIUM NUCLEUS
TRITIUM NUCLEUS (ANOTHER FORM OF HYDROGEN)
ENERGY
NEUTRON

------ WHAT HAPPENED WHEN THE UNIVERSE WAS CREATED? **SIDE ROAD** TO ROAD STOP 4 ------→

------ IS THERE AN ENDLESS SUPPLY OF ENERGY? **SIDE ROAD** TO ROAD STOP 5 ------→

ROAD STOP 4

What is the big bang – and what does it have to do with energy?

Most scientists believe that the universe was formed from an enormous explosion, called the big bang. Scientists can trace the history of energy back to the big bang.

Visible Proof SPOT

According to the big bang theory, the universe is constantly expanding. To see how it grows, stick some star shapes on a deflated balloon. When you blow up the balloon, it expands and the stars spread out – just as the space between real galaxies moves apart.

Detour
SOME SCIENTISTS BELIEVE THAT ONE DAY THE UNIVERSE WILL COLLAPSE. FIND OUT MORE IN **ROAD STOP 5**.

1 ▶ SOLAR SYSTEM

The solar system consists of the sun and all the planets, moons, and other heavenly bodies that circle, or orbit, around it.

2 ▶ GALAXY

A galaxy is a large group of stars, dust, and gas. The sun is one of more than 100 billion stars that make up the Milky Way Galaxy.

3 ▶ UNIVERSE

The universe is the whole of space and everything in it. It contains billions of galaxies, most of them grouped into clusters.

▶ FOLLOW THE **HIGHWAY** AND FIND OUT ABOUT MUSCLE POWER.

EVIDENCE FOR THE BIG BANG
In 1965, American scientists Arno Penzias and Robert Wilson discovered a faint radio signal in space. The signal appeared to fill the entire universe. Scientists believed that it was caused by heat left over from the massive radiation created by the big bang.

WHAT IS THE BIG BANG?

※ Many scientists believe that all energy, matter, space, and time was once squeezed tightly together in a single, tiny nucleus. This concentrated point is called a singularity.

※ About 15 billion years ago, this nucleus exploded, creating the big bang. Scientists do not know what caused the explosion.

※ The nucleus released a tremendous amount of radiation, which expanded rapidly. Within minutes, it created basic pieces of matter, such as atoms of helium.

※ As the universe expanded, it cooled. Hydrogen and other light gases formed. Over millions of years, the gases came together and produced stars, galaxies, and planets.

ROAD STOP 5

What is the big crunch?

Detour
WE NEED TO CUT DOWN THE AMOUNT OF ENERGY WE USE. DETOUR TO **ROAD STOP 26**.

Some scientists think that in billions of years the universe will collapse in a massive explosion of energy, called the big crunch.

THE BIG CRUNCH
The big crunch theory is like the big bang in reverse. It says that all energy will eventually condense back into one, tiny particle.

1 In about 70 billion years' time, the universe will stop expanding and begin to shrink, or contract.

2 The big crunch will occur as all the galaxies come together and collide in a single, incredibly hot singularity.

3 A whole new universe will be created when the singularity explodes again, in another big bang.

All living things need energy to survive.

Select
SIDE ROAD

Energy
FOR LIFE

6 Why do living things need energy?

7 How do plants use the sun's energy?

8 Where do people get their energy from?
(WHAT IS A FOOD CHAIN?)

9 What is muscle energy?

10 What effect does energy have on our bodies?
(CAN PEOPLE CONTROL THE ENERGY IN THEIR BODIES?)

Beware!
If the body does not use all the energy it receives from food, it may convert some of it into fat!

ROAD STOP 6

Why do living things need energy?

Detour
WHAT EFFECT CAN ENERGY HAVE ON THE HUMAN BODY? LEAP TO ROAD STOP 10.

All living things need energy to survive. They use energy to stay warm, communicate, move about, catch food, and build homes.

SCIENTISTS HAVE IDENTIFIED MORE THAN ONE AND A HALF MILLION KINDS OF ANIMALS IN THE WORLD. EACH CREATURE NEEDS AND USES ENERGY.

❋ A KANGAROO NEEDS ENERGY TO JUMP.

❋ A LION NEEDS ENERGY TO ROAR.

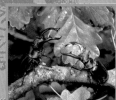
❋ A STAG BEETLE NEEDS ENERGY TO FIGHT.

FOLLOW THE **HIGHWAY**.

ROAD STOP 7

How do plants use the sun's energy?

MAKING FOOD

Chlorophyll traps light energy. The energy drives a chain of chemical reactions, in which water and carbon dioxide gas combine to produce a simple sugar (glucose) and oxygen. The plant uses the sugar for food and releases the oxygen into the air.

LIGHT ENERGY FROM THE SUN
CARBON DIOXIDE FROM THE AIR
CELLS IN THE LEAF CONTAIN A GREEN PIGMENT CALLED CHLOROPHYLL
WATER FROM THE SOIL

Green plants capture the energy in sunlight to make food. This process is called photosynthesis, which means "putting together by light."

HOW DO WE GET ENERGY FROM FOOD? **SIDE ROAD** TO ROAD STOP 8
HOW DO PEOPLE USE THEIR ENERGY EFFICIENTLY? **SIDE ROAD** TO ROAD STOP 9
HOW MUCH ENERGY DO WE NEED? **SIDE ROAD** TO ROAD STOP 10

ROAD STOP

Where do people get their energy?

Detour

SIDE ROAD TO ROAD STOP 8

WHAT ELSE STORES ENERGY FROM THE SUN? DETOUR TO **ROAD STOP 28**.

Unlike plants, most animals cannot trap the energy in sunlight to make their own energy. They have to use energy stored elsewhere. Humans get their energy from the food they eat.

TURNING STORED ENERGY IN FOOD INTO POTENTIAL ENERGY IN PEOPLE

1 You eat food, such as fruit, vegetables, fish, and meat.

2 Your body breaks down, or digests, the food.

3 As the digested food mixes with oxygen in your body, it burns. The food's chemical energy is released.

4 Any energy that your body does not use immediately is stored in your cells as potential energy.

PRODUCERS

GREEN PLANTS ARE KNOWN AS PRODUCERS. THEY MAKE, OR PRODUCE, THEIR OWN FOOD FROM NON LIVING MATTER (SUNLIGHT).

CONSUMERS

CONSUMERS ARE LIVING THINGS THAT EAT, OR CONSUME, OTHER LIVING THINGS. ALMOST ALL ANIMALS ARE CONSUMERS.

Human consumers obtain their energy from the chemical energy stored in other living things.

Detour

HOW DOES ENERGY KEEP US WARM? LEAP TO **ROAD STOP 23**.

FOLLOW THE **HIGHWAY** TO DISCOVER WHICH SCIENTISTS EXPLORED ENERGY AND EXPLAINED IT TO US.

ROAD STOP

What is muscle energy?

Energy from food may be stored in the body's muscles as a form of potential energy, called muscle energy. We use muscle energy to perform tasks. Tools and other machines help us to use this energy more efficiently.

Detour

FIND OUT ABOUT USEFUL MACHINES THAT USE DIFFERENT FORMS OF ENERGY. LEAP TO **ROAD STOP 24**.

SOME LIVING THINGS ARE PRODUCERS, AND OTHERS ARE CONSUMERS. SIDE ROAD TO ROAD STOP 8

SIDE ROAD TO ROAD STOP 9

The Egyptian pyramids were built about 4,500 years ago, using the muscle energy of thousands of men. Sledges made it easier for workers to drag the blocks to the pyramid.

☼ USING MUSCLE ENERGY TO FARM THE LAND For thousands of years, farmers have used plows to break up the soil. The first plows were just forked branches that people pulled by themselves. Later, they learned to use the muscle energy of animals by hitching oxen and horses to plows.

☼ USING MUSCLE ENERGY IN TRANSPORTATION Rowing a boat with oars uses muscle energy. Before people learned to build sails to harness the energy of the wind, boats were powered by oarsmen. Sometimes, thousands of slaves were forced to row huge Roman ships, called galleys.

10

ENERGY CAN HAVE GOOD OR BAD EFFECTS ON THE BODY. **SIDE ROAD** TO ROAD STOP **10**

ENERGY IN FOOD

Different foods store and release different amounts of energy. Food contains three main groups of energy-giving chemicals, or nutrients: carbohydrates, proteins, and fats.

1 CARBOHYDRATES
ROLE: main source of energy for plants and animals.
FOUND: in sugary foods, such as jam, honey, and candy; in starchy foods, such as bread, rice, and potatoes.
ENERGY: 1 gram of carbohydrate = about 4 kilocalories.

2 PROTEINS
ROLE: help the body to grow and repair itself.
FOUND: in many foods, including meat, fish, cheese, eggs, milk, beans, and nuts.
ENERGY: 1 gram of protein = about 4 kilocalories.

3 FATS
ROLE: provide energy and insulate against heat loss.
FOUND: in foods such as butter, cream, cheese, yogurt, milk, nuts, and oil.
ENERGY: 1 gram of fat = about 9 kilocalories.

FOLLOW THE PATH TO THE POINT OF INTEREST TO FIND OUT ABOUT FOOD CHAINS.

POINT of Interest

Energy is passed on along a chain.

What is a food chain?

A food chain is a series of living things, each of which is eaten by the next. A food chain shows how food, and therefore energy, passes from one living thing to another.

A SIMPLE FOOD CHAIN

PRODUCER → **PRIMARY CONSUMER** → **SECONDARY CONSUMER**

Green plants, such as grass, combine energy in sunlight with energy from water and nutrients in the soil, to make food.

A plant-eating animal (herbivore), such as a rabbit, eats the grass. Some of the sun's energy stored in the grass passes onto the rabbit.

A meat-eating animal (carnivore), such as a fox, eats the rabbit. Some of the energy from the rabbit passes on to the fox.

DECOMPOSER Decomposers, such as bacteria and fungi, break down the remains of the dead grass, rabbits, and foxes into simple nutrients. These return to the soil and are used as energy by other plants. At this and every stage in a food chain, some energy turns to heat and most of it escapes into the air.

Visible Proof SPOT

A typical food chain joins together three or four different living things. Each living link in the chain is a source of food for the next one. Put the pictures below in the correct order to make a food chain.

ANSWER: SUNSHINE PROVIDES ENERGY FOR THE CABBAGES TO MAKE THEIR OWN FOOD; CATERPILLARS EAT CABBAGE LEAVES; SPARROWS EAT CATERPILLARS; SPARROWS ARE EATEN BY THE CAT.

---- FOLLOW THE **HIGHWAY**. ---->

ROAD STOP 10

What effect does energy have on our bodies?

THE BODY USES ENERGY TO:
- keep itself warm.
- make the muscles work.
- enable the brain to think.
- enable the body to grow, repair itself, and stay healthy.
- keep the heart beating.

The amount of energy people need each day depends on their age, sex, and what they are doing. Energy is essential for life, but too much energy can cause health problems.

HOW MUCH ENERGY DO YOU NEED?

To stay healthy and active, different people need different amounts of energy each day.

Toddler (1-3 years)	1,300 kcal	(5,500 kJ)
Child (7-10 years)	2,000 kcal	(8,400 kJ)
Girl (11-14 years)	2,200 kcal	(9,200 kJ)
Boy (11-14 years)	2,500 kcal	(10,500 kJ)
Woman (25-50 years)	2,200 kcal	(9,200 kJ)
Man (25-50 years)	2,900 kcal	(12,200 kJ)

HEALTHY BODY
Chemical energy released from food helps this athlete to build strong muscles for running and competing. She is at the peak of physical fitness.

UNHEALTHY BODY
This person is severely overweight. He is taking in more energy, through eating food, than he uses. He stores the energy from the excess calories in his body as fat.

DOES ENERGY FLOW THROUGH THE BODY? FOLLOW THE PATH TO **THE POINT OF INTEREST** TO FIND OUT.

POINT of Interest

Energy that causes illness.

Can people control the energy in their bodies?

Acupuncture and reflexology are two ancient forms of healing. People who practice them believe that they can tap into and control the energy that runs through the body.

ACUPUNCTURE

Acupuncture is an ancient Chinese method of stopping pain and treating disease. An acupuncturist inserts thin needles at various points on a patient's body. This is said to restore the balance of energy, called chi, that is believed to flow along 14 paths, or meridians, in the body. An imbalance of chi is said to cause ill-health.

FOLLOW THE **HIGHWAY** AND MEET THE PEOPLE WITH THE ANSWERS.

REFLEXOLOGY

Reflexologists believe that energy exists in 10 zones in the body, each beginning in the toes and ending in the fingers. They say that when energy pathways become blocked, a person becomes ill. A reflexologist massages the patient's feet and hands to clear the blockages and restore the flow of energy to the body.

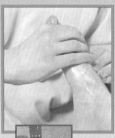

 Visible Proof **SPOT**

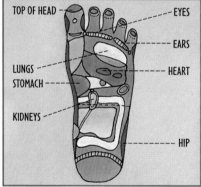

TOP OF HEAD — EYES
— EARS
LUNGS
STOMACH — HEART
KIDNEYS
— HIP

A reflexologist applies pressure to specific areas of the feet and hands to stimulate corresponding parts of the body. Here is a reflexology chart of the foot. The next time you have a headache, try massaging your big toe. Look carefully at the chart to find out which region of your foot you should massage if you have a stomach ache.

NO NEEDLES
Acupressure is similar to acupuncture but, instead of needles, pressure is applied to the body using the fingers and hands.

Some important scientists have helped to explain what energy is all about.

Select

YOUR SIDE ROAD

LOOKING FOR

Answers

11 Who first studied energy?

12 Which scientist experimented with potential and kinetic energy?
(WHO ELSE EXPLORED ENERGY?)

13 What is the quantum theory?

14 Why was Albert Einstein's work so important?
(WHAT DOES $E = mc^2$ MEAN?)

ROAD STOP 11

Who first studied energy?

Detour
WHEN DID SCIENTISTS FIRST SPLIT AN ATOM? LEAP TO **ROAD STOP 14**.

The ancient Greeks were among the first people to think seriously about energy and its effects on matter.

ANAXAGORAS (500?-428 B.C.)

Anaxagoras was a great thinker, or philosopher, who believed in a powerful force called "nous," meaning the mind. He thought that the universe was created through the action of nous on tiny particles of matter. He called these particles "seeds."

NOUS = SIMILAR TO WHAT WE THINK OF AS ENERGY.
SEEDS = SIMILAR TO WHAT WE THINK OF AS ATOMS.

DEMOCRITUS (460?-370? B.C.) & LEUCIPPUS (400's B.C.)

These two philosophers developed the theory of atomism. This stated that everything was made of invisible particles called atoms, which could not be divided any further, unlike Anaxagoras's seeds, which could be divided infinitely.

THE WORD "ATOM" COMES FROM THE GREEK WORD "ATOMOS," MEANING UNCUTTABLE.

ARISTOTLE (384-322 B.C.)

This philosopher and scientist saw no point in investigating the energy that moved objects. He believed that things moved simply because it was their nature to do so. For example, a rock falls because it is heavy and a kangaroo jumps because that is the natural thing for it to do. This idea was accepted for almost 2,000 years.

Visible Proof SPOT

Atoms move to and fro, or vibrate, all the time. Put a few drops of food coloring in a glass of water. The drops sink to the bottom. Leave the glass undisturbed and return to it in about an hour. The water is now completely colored. The movement of the atoms in the water has stirred in the coloring.

FOLLOW THE **HIGHWAY**.

ROAD STOP 12

Which scientist first experimented with potential and kinetic energy?

The Italian scientist Galileo Galilei (1564-1642) was one of the first people to investigate the properties of moving objects.

CHANGING ENERGY
Galileo carried out many experiments with objects falling through the air and rolling down slopes. He was aware of the idea of potential and kinetic energy, and realized that one could turn into the other.

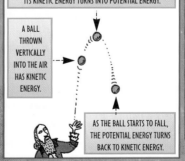

AT ITS HIGHEST POINT, THE BALL STOPS MOVING BRIEFLY AND ITS KINETIC ENERGY TURNS INTO POTENTIAL ENERGY.

A BALL THROWN VERTICALLY INTO THE AIR HAS KINETIC ENERGY.

AS THE BALL STARTS TO FALL, THE POTENTIAL ENERGY TURNS BACK TO KINETIC ENERGY.

FALLING BODIES
Galileo believed that objects would fall at the same speed, regardless of their size, shape, or weight. One story is that he proved his theory by dropping two cannon balls, one small and one large, from the top of the Leaning Tower of Pisa. They landed on the ground at the same time.

FOLLOW THE PATH TO THE **POINT OF INTEREST**.

WHO DISCOVERED THAT ENERGY EXISTED IN TINY PACKETS? **SIDE ROAD** TO ROAD STOP 13
WHO WAS ALBERT EINSTEIN? **SIDE ROAD** TO ROAD STOP 14

POINT of Interest
Making sense of energy.

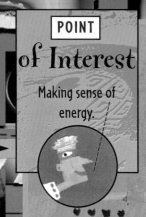

Who else explored energy?

Detour
CAN SOMETHING KEEP MOVING FOREVER WITHOUT A FORCE TO KEEP IT GOING? LEAP TO ROAD STOP 30.

Scientists during the 1600's and 1700's made additional discoveries about energy.

GASPARD DE CORIOLIS (1792-1843)
This Frenchman devised a formula to calculate the kinetic energy of a moving object.

THE FORMULA: KE = $1/2 \times m \times v^2$
KE = KINETIC ENERGY m = MASS OF THE OBJECT
v^2 = VELOCITY (SPEED IN A PARTICULAR DIRECTION) OF THE OBJECT MULTIPLIED BY ITSELF

SIR ISAAC NEWTON (1642-1727)
This English scientist formulated three laws of motion to explain how objects move. He also carried out experiments with light and showed that white light is made up of different colors.

BENJAMIN THOMPSON (1753-1814)
This scientist, born in America, concluded that heat is a form of motion, not an invisible fluid (the common belief at the time).

→ FOLLOW THE **HIGHWAY** AND FIND OUT HOW ENERGY CHANGES FROM ONE FORM TO ANOTHER.

ROAD STOP 13

What is the quantum theory?

Detour
HOW DO WE USE HEAT ENERGY IN THE HOME? LEAP TO **ROAD STOP 23**.

This theory states that light, heat, and other radiant energy is made up of a stream of tiny particles. Each particle, or quantum, carries a certain amount of energy.

Who came up with this theory? In 1900, the German scientist Max Planck (1858-1947) introduced the quantum theory to explain how hot objects give off light. He suggested that objects absorb and emit radiant energy not in waves, but in packets of energy, or quanta.

VARYING ENERGIES
Different colors of light contain quanta carrying different amounts of energy. For example, a quantum of blue light has about twice as much energy as a quantum of red light.

A horseshoe may be heated until it is white-hot. Its atoms collide violently, releasing quanta of light.

Visible Proof SPOT

To see how heat makes an object give off light, ask an adult to turn on a ring of an electric stove. As the ring heats up, it turns orange, then red. When the ring is turned off, it cools down and returns to its normal color.

ROAD STOP 14

Why was Albert Einstein's work so important?

Albert Einstein (1879-1955) was a German-born, American scientist whose work completely changed people's understanding of atoms, mass, and energy.

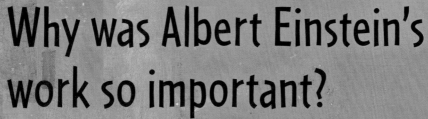

A solar-powered calculator uses the photoelectric effect to produce an electric current to display the numbers.

WHAT DID EINSTEIN DISCOVER?

In 1905, while working in the Swiss Patent Office in Bern, Einstein published three groundbreaking scientific papers:
PAPER 1 = Explained the photoelectric effect.
PAPER 2 = Proved the existence of atoms.
PAPER 3 = Showed that energy has mass.

THE PHOTOELECTRIC EFFECT

1. Quanta of light, called photons, strike the metal atoms.
2. The energy in the photons transfers to the electrons, which then shoot out.
3. The ejected electrons may form an electric current.

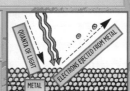

When a bright beam of light hits a metal, it can produce a weak electric current. This is called the photoelectric effect. Einstein developed Planck's quantum theory to explain it.

Detour

FIND OUT MORE ABOUT ELECTRICITY IN **ROAD STOP 18**.

FOLLOW THE PATH TO THE **POINT OF INTEREST** TO LEARN ABOUT A FAMOUS EQUATION.

FOLLOW THE **HIGHWAY**.

POINT of Interest

The most famous formula of all time!

What does $E = mc^2$ mean?

Einstein developed this formula to describe the relationship between the mass of a given amount of matter and its energy.

THE FAMOUS FORMULA

$E = mc^2$ (E EQUALS M TIMES C-SQUARED)

E represents energy **m** represents mass

c^2 is the velocity of light multiplied by itself

WHAT THE FORMULA MEANS

※ Einstein's formula states that mass (m) can be changed into energy (E). Since the velocity of light (c) is a large number – 299,792 kilometers per second – a tiny amount of mass can create a vast amount of energy.

※ A mass of one gram can produce enough electricity to keep a light bulb burning continuously for more than 28,500 years.

Detour

WHEN ATOMS ARE SPLIT, THEY PRODUCE NUCLEAR ENERGY. FIND OUT MORE BY LEAPING TO **ROAD STOP 29**.

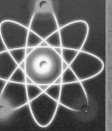

PROVING EINSTEIN'S THEORY
Scientists were not able to change mass into energy until the 1930's, when they discovered how to split an atom. They noticed that during a reaction, some of the atom's mass disappeared; at the same time, energy was produced. They concluded that the missing mass had been changed into this energy.

※ Einstein proved that the mass and energy of an object increase the faster it moves. A spaceship cannot travel at the speed of light because it would need a never-ending amount of energy.

Energy rarely stays in the same form.

Select
YOUR SIDE ROAD

Energy IS TRANSFORMED

15 When is energy transformed?

16 What happens when potential energy is transformed?

17 Does the transformation of energy change the total amount of energy?

Wow!
In theory, the energy that was released by the big bang still exists in the universe today!

ROAD STOP 15

When is energy transformed?

Every form of energy can be converted into a different form. Energy changes form whenever work is done. We call this process of change transformation of energy.

ENERGY MUST CHANGE FORM SEVERAL TIMES FOR YOU TO BE ABLE TO LIFT A BALL.

FOLLOW THE **HIGHWAY** TO FIND OUT ABOUT ELECTRICITY.

ROAD STOP 16

What happens when potential energy is transformed?

Detour
DISCOVER HOW A BATTERY RELEASES ITS POTENTIAL ENERGY. LEAP TO ROAD STOP 19.

Potential energy is stored energy. When it is converted into a different type of energy, it can be used in a powerful way, to make things happen.

1 When you lift an object, such as a tray, into the air, you convert the muscle energy in your arms to kinetic energy.

2 Potential energy is created whenever an object moves against a force acting on it. Lifting the tray makes it work against gravity, a natural force that pulls objects towards the ground.

3 The tray has now gained potential energy. If you let go of the tray, it falls to the floor because it has been given potential energy.

WHAT HAPPENS TO ENERGY WHEN IT IS CHANGED? **SIDE ROAD** TO ROAD STOP 17

EXAMINE THIS SERIES OF ENERGY TRANSFORMATIONS

1 The sun radiates light energy.

2 An apple tree absorbs light energy. The tree stores some as chemical energy in the fruit it produces.

3 When you eat an apple its chemical energy is released into your body.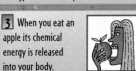

4 Your muscles turn some of the chemical energy into kinetic energy when you move your leg.

5 When you kick a ball, you transfer kinetic energy to the ball, sending it flying through the air.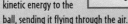

Visible Proof SPOT

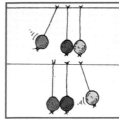

Thread three heavy beads onto pieces of string and hang them so that the beads touch. Lift one of the end beads and drop it. Its kinetic energy travels through the row and pushes up the bead at the other end. The beads keep swinging until all the energy has been converted to sound (the clicking) and heat (created by the beads colliding). The beads also lose energy to friction as they move through the air.

DRAMATIC TRANSFORMATIONS

During a thunderstorm, electrical energy in the atmosphere is transformed into light (lightning), sound (thunder), and heat energy.

Explosives store chemical energy that can be released quickly. When a firework is lit, its chemical energy is converted into light (the firework display), sound (the bang), kinetic energy (which makes the firework soar into the sky), and heat.

Detour

HOW IS CHEMICAL ENERGY TURNED INTO ELECTRICAL ENERGY? LEAP TO ROAD STOP 19.

--- FOLLOW THE **HIGHWAY**. --->

ROAD STOP 17

Does the transformation of energy change the total amount of energy?

Energy changes from one form to another, but the total amount of energy in the universe remains the same. Energy is said to be conserved.

Visible Proof SPOT

Place an elastic band around your thumb and finger. Wrap a small piece of paper around the far side of the band, then pull it toward you. The stretched band gains potential energy. Aim the band at a wall and release the paper. The potential energy of the band changes to kinetic energy that launches the paper through the air.

SIDE ROAD TO ROAD STOP 17

AN IMPORTANT LAW

The law of conservation of energy states that the total amount of energy in the universe stays the same. Energy cannot be created or destroyed, it can only be transformed. Energy that seems to disappear is actually converted to another form of energy, such as sound or heat.

NUCLEAR ENERGY

During a nuclear reaction, matter is converted into energy. This suggests that energy has been created. But Albert Einstein showed that matter is a form of energy, so the total amount of energy remains unchanged and the law of conservation of energy still applies.

Electricity – energy that powers our planet.

Select
YOUR SIDE ROAD

ALL ABOUT
Electricity

18 What is electricity?
(WHO FIRST INVESTIGATED ELECTRICITY?)

19 What is an electric circuit?
(HOW DOES A BATTERY WORK?)

20 How else do we generate electricity?
(HOW DOES ELECTRICITY MOVE FROM ONE PLACE TO ANOTHER?)

21 Why is electricity essential for life?

Warning!
Never experiment with electrical outlets – they are extremely dangerous.

ROAD STOP 18

What is electricity?

Electricity is a form of energy associated with the movement of electrons. People began to learn about electricity as early as the 500's B.C.

BALANCED CHARGES
Electrons circle around the nucleus of an atom. Electrons carry negative electric charges, and the nucleus carries equivalent positive charges. The charges balance each other out, so the atom is electrically neutral.

AN ATOM BECOMES AN ION
Sometimes, an atom loses or gains one or more electrons. If it gains an electron, the atom takes on a negative electric charge. If it loses an electron, the atom takes on a positive charge. A charged atom is called an ion.

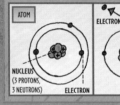

ATOM — NUCLEUS (3 PROTONS, 3 NEUTRONS) — ELECTRON — ION

MAKING AN ELECTRIC CURRENT

An electric current is a flow of electric charges through a material. Electricity flows well through metals because they contain a large number of electrons that can move freely between the atoms.

FOLLOW THE **HIGHWAY** TO FIND OUT HOW WE USE ENERGY.

ROAD STOP 19

What is an electric circuit?

Electric charges flow in an electric current. The path that an electric current follows is called an electric circuit.

LOOKING AT CIRCUITS
※ In these circuit diagrams, an electric current flows from a battery (the energy source) to a light bulb (the electric device).
※ When the switch is open, there is a gap in the circuit and no current flows.
※ As the current passes through the bulb, most of the electric energy is converted into light and heat energy.

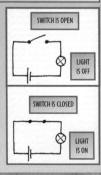

SWITCH IS OPEN — LIGHT IS OFF
SWITCH IS CLOSED — LIGHT IS ON

Visible Proof SPOT

Static electricity is a type of electricity that does not move. It can be produced by rubbing certain things together. Run a comb through your hair several times. The comb moves electrons to your hair and gains a positive charge. When you hold the comb above small pieces of paper, the comb's positive charge attracts the paper and makes it stick to the comb.

FOLLOW THE PATH TO THE POINT OF INTEREST TO FIND OUT WHO FIRST EXPLORED ELECTRICITY.

POINT of Interest
Explaining the mystery of electricity.

Who first investigated electricity?

The ancient Greeks were the first to experiment with electricity.

THALES OF MILETUS (625?-546? B.C.)

This Greek philosopher carried out simple experiments to explore the nature of electricity. He discovered that when he rubbed a natural material called amber with cloth, it became electrically charged and attracted light objects, such as feathers. The word "electricity" comes from "elektron," the Greek word for amber.

Amber becomes electrically charged when it is rubbed with a cloth.

TO FIND OUT ABOUT BATTERIES, FOLLOW THE PATH TO THE POINT OF INTEREST.

CIRCUIT SYMBOLS
Special symbols represent parts of a circuit diagram.

- CONNECTING WIRE
- BATTERY
- SWITCH
- LIGHT BULB

POINT of Interest
Find out about portable electricity.

How does a battery work?

A battery is a device that stores chemical energy and turns it into electrical energy.

WHO INVENTED THE BATTERY?

An Italian scientist called Count Alessandro Volta (1745-1827) invented the first practical battery in the late 1790's. He gave his name to the unit of electric measurement, the volt.

INSIDE A BATTERY

A battery contains an electrolyte, a liquid or paste of chemicals, which produces an electric current.

- POSITIVE ELECTRODE (+) — Positive electric charges build up at this electrode.
- ELECTROLYTE. Here chemicals react with one another and split into ions, producing an electric current.
- ELECTRIC CURRENT FLOWS OUT OF THE NEGATIVE ELECTRODE, PASSES THROUGH THE BATTERY AND RETURNS THROUGH THE POSITIVE ELECTRODE
- NEGATIVE ELECTRODE (−) — Negative electric charges build up at this electrode.

Visible Proof SPOT

Make a battery from a lemon. Carefully cut two slits in the lemon and push a piece of aluminum foil in one slit and a copper coin in the other. Make a circuit by touching the foil and coin to your tongue. You should feel a tingling sensation. The lemon juice (electrolyte) reacts with the two metals (electrodes) to produce a weak electric current.

Detour
WHAT IS FUEL? LEAP TO ROAD STOP 23.

HOW IS ELECTRICITY MOVED FROM ONE PLACE TO ANOTHER? SIDE ROAD TO ROAD STOP 20

WHY IS ELECTRICITY CRUCIAL TO LIFE? SIDE ROAD TO ROAD STOP 21

ROAD STOP 20

How else do we generate electricity?

Detour
HOW DOES A NUCLEAR POWER STATION WORK? LEAP TO **ROAD STOP 29**.

Most of the electricity we use every day is made, or generated, in huge power stations. Here large machines, called electric generators, turn kinetic energy into electricity.

This hydroelectric dam project in Venezuela uses the power of water falling from a reservoir to drive a generator.

LOOKING BACK
In 1831, the English scientist Michael Faraday (1791-1867) found that passing a magnet through a coil of copper wire made an electric current flow through the wire. This idea led to the invention of the electric generator.

PRODUCING ELECTRICITY
Many power stations use steam-driven turbines to power generators.

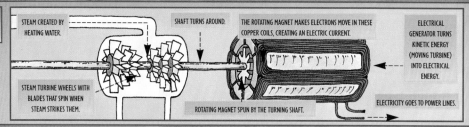

STEAM CREATED BY HEATING WATER. — SHAFT TURNS AROUND. — THE ROTATING MAGNET MAKES ELECTRONS MOVE IN THESE COPPER COILS, CREATING AN ELECTRIC CURRENT. — ELECTRICAL GENERATOR TURNS KINETIC ENERGY (MOVING TURBINE) INTO ELECTRICAL ENERGY.

STEAM TURBINE WHEELS WITH BLADES THAT SPIN WHEN STEAM STRIKES THEM. — ROTATING MAGNET SPUN BY THE TURNING SHAFT. — ELECTRICITY GOES TO POWER LINES.

FOLLOW THE **HIGHWAY** TO DISCOVER HOW ENERGY HAS BEEN USED TO TRANSFORM THE WORLD.

ROAD STOP 21

Why is electricity essential for life?

Visible Proof SPOT

Stack a row of dominoes on a flat surface. When you push over the first domino, the others fall over, one after another. This is how a message travels along a nerve. One neuron fires an impulse to the next neuron, which fires an impulse to the next neuron, and so on.

All animals, including humans, have a small amount of electricity in their bodies. It helps to drive vital body processes. Some animals can generate powerful blasts of electricity.

Nerve signals

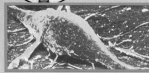

Nerves throughout the body collect information from the senses and send it to the brain. The messages are transmitted as tiny electrical signals, or impulses. The brain interprets the signals and sends out its own impulses to the muscles and organs.

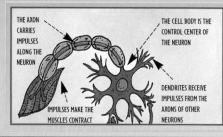

THE AXON CARRIES IMPULSES ALONG THE NEURON — THE CELL BODY IS THE CONTROL CENTER OF THE NEURON — DENDRITES RECEIVE IMPULSES FROM THE AXONS OF OTHER NEURONS — IMPULSES MAKE THE MUSCLES CONTRACT

A NERVE IS MADE UP OF NERVE CELLS, OR NEURONS, LINKED TOGETHER.

Visible Proof SPOT

Make a heat-driven turbine. Wrap modeling clay round the blunt end of a knitting needle and wedge strips of cardboard into it at angles. Stick the needle in a spool on a tray. Ask an adult to light small candles underneath. As the hot air rises, it makes the turbine spin. Blow out the candles once you have finished.

Detour

FIND OUT MORE ABOUT ALTERNATIVE WAYS OF GENERATING ENERGY IN **ROAD STOP 27**.

HOW IS ELECTRICITY DISTRIBUTED? FOLLOW THE PATH TO THE **POINT OF INTEREST**.

POINT of Interest

Electricity on the move.

How does electricity move from one place to another?

Power stations send out electricity to places where it is needed, along overhead wires, called transmission or power lines. Some lines are buried underground or underwater.

FROM POWER STATION TO ELECTRICAL OUTLET

1 THE POWER STATION SENDS THE ELECTRICITY PRODUCED BY THE GENERATOR ALONG CABLES TO A DEVICE CALLED A STEP-UP TRANSFORMER.	2 THIS TRANSFORMER INCREASES THE VOLTAGE OF THE CURRENT SO THAT IT CAN BE SENT LONG DISTANCES.	3 AT A SUBSTATION, A STEP-DOWN TRANSFORMER REDUCES THE VOLTAGE OF THE CURRENT. LARGE FACTORIES TAKE POWER AT HIGH VOLTAGES.	4 AT A SMALLER SUBSTATION A STEP-DOWN TRANSFORMER FURTHER REDUCES THE OUTPUT. POWER LINES CARRY THE ELECTRICITY TO ELECTRICAL OUTLETS IN HOMES AND OFFICES.
ABOVE CABLES HAVE UP TO 22,000 VOLTS	ABOVE CABLES HAVE UP TO 765,000 VOLTS	ABOVE CABLES HAVE 2,000 TO 34,500 VOLTS	ABOVE CABLES HAVE 110 TO 240 VOLTS

---- FOLLOW THE **HIGHWAY**. ---->

2 A human heartbeat

About every second, electrical signals spread through the muscles of the heart. These signals make the heart beat. A machine called an electrocardiograph can detect and record this electrical activity.

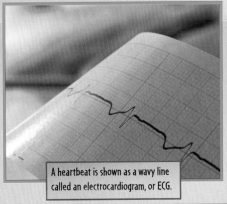

A heartbeat is shown as a wavy line called an electrocardiogram, or ECG.

BRINGING MONSTERS TO LIFE

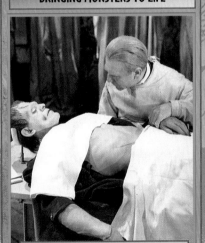

In the story Frankenstein, a scientist makes a monster he has created come to life by passing an electric current through its body.

ELECTRIC FISH

Some fish have muscles in their bodies that can generate electricity. The fish send powerful electric shocks through the water to protect themselves from enemies and to stun or kill prey.

※ An electric ray, above, has muscles on both sides of its head that give off an electric current.
※ An electric eel can generate up to 650 volts of electricity. This is enough to stun a person.

21

We've looked at what energy is and how it can be transformed – now let's use it!

Arrival
AT DESTINATION
HARNESSING Energy

FOLLOW THE **HIGHWAY**.

Think!
Some fuels are made from the remains of plants and animals that lived on the earth millions of years ago.

Energy can be used to make the world more comfortable and interesting!

Select
YOUR SIDE ROAD
Using ENERGY

22 What forms of energy did our ancestors use?

23 How do we use energy to keep warm?
(WHAT ARE FOSSIL FUELS?)

24 How does energy power machines?
(HOW DOES ENERGY HELP US TO COMMUNICATE?)

25 How much energy is used around the world?

ROAD STOP 22

What forms of energy did our ancestors use?

In the past, people harnessed the natural energy around them. They used the kinetic energy of wind and moving water to drive machinery. They also used heat and light energy from fires.

Detour

SEE HOW THE WIND CAN BE USED TO GENERATE ELECTRICITY. LEAP TO **ROAD STOP 27**.

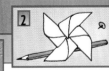

1. Wind power

※ **3000 B.C.** The ancient Egyptians invented the sail, which they discovered could catch the wind and propel their boats. They used a large, rectangular sail made of cloth.

※ **600's** The first windmills were probably used in Persia, now called Iran. The wind turned the mills and the vanes turned stones that ground grain.

Visible Proof SPOT

Make a simple windmill. Take a piece of paper about 20 in. (15 cm) square and carefully cut each corner to within 1 in. (2.5 cm) of the center. Lift two opposite corners to the center and use a drawing pin to tack the corners to a pencil. Now hold your windmill in the wind and watch it spin!

FOLLOW THE **HIGHWAY** AND DISCOVER HOW ENERGY IS USED TODAY.

---

2. Water power

※ **100's B.C.** Water wheels were used in several parts of the world to grind grain. As the wheel turned, it changed the kinetic energy of falling or flowing water into mechanical energy.

※ **1700's** Factories and mills used large water wheels to run machinery. This power was not reliable. In a dry season water levels dropped, leaving the factories with reduced or no power.

3. Fire power

WHAT IS FIRE?

Fire is the heat and light that comes from a burning object. A fire is caused by a combination of three things: heat, oxygen, and fuel.

WHAT IS FUEL?

Fuel is any material that releases useful energy, either by burning with oxygen or by atomic means.

※ **600,000 B.C.** Prehistoric people, called homo erectus, were probably the first human beings to master fire. The remains of burned bones and ash found in caves suggest that they used fire to cook food. They also used it to keep warm and provide light.

Here food is being cooked in clay pots over an open fire.

Detour

WHAT ARE FOSSIL FUELS? LEAP TO **ROAD STOP 23**.

※ **1500 B.C.** The Egyptians used the heat from fire to make glass vessels. Glass is made mainly from sand, mixed with other materials. A glassmaker heats the mixture until it melts. When it cools, it sets as glass.

WHAT IS A HYPOCAUST? SIDE ROAD TO ROAD STOP 23

HOW IS ENERGY USED TO COOK FOOD? SIDE ROAD TO ROAD STOP 24

DO RICH COUNTRIES USE MORE ENERGY THAN POOR COUNTRIES? SIDE ROAD TO ROAD STOP 25

ROAD STOP

23

How do we use energy to keep warm?

EXPLORING HEAT ENERGY
During the 1840's, the English scientist James Joule and the German scientists Julius von Mayer (1814-1878) and Hermann von Helmholtz (1821-1894) independently carried out experiments proving that heat is a form of energy.

HEAT
All things are made of atoms that are always moving. The faster the atoms move, the higher the level of energy inside the object (called internal energy) and the higher the temperature.

TEMPERATURE
Temperature is a measure of an object's internal energy. The more internal energy, the hotter an object is.

Heat is a form of energy. Eating food releases stored chemical energy that our bodies turn to heat, to keep us alive. We can also create heat from friction and by burning fuel.

HOW DO WE CREATE HEAT ENERGY?

1 COMBUSTION
Flammable materials can catch fire, or combust. As the materials burn, they produce light and heat.

2 FRICTION
Friction is a force produced when one object rubs against another. As objects rub together, their atoms vibrate and generate heat.

3 USING ANOTHER ENERGY SOURCE
Appliances such as electric heaters, kettles, and stoves use electricity to create heat.

HEATING THE HOME

THEN HYPOCAUST
The ancient Romans invented this form of heating about 100 B.C. Fires produced hot air that circulated under the floors and through the walls of a building, heating the rooms as a result.

NOW CENTRAL HEATING
In one form of central heating, a boiler, powered by gas, electricity, or oil, heats water. It sends the hot water to a radiator, a set of pipes or tubes that transfers the heat to the air in a room.

FOLLOW THE **HIGHWAY** TO DISCOVER HOW PEOPLE ARE EXPLORING NEW WAYS OF MEETING THE WORLD'S ENERGY REQUIREMENTS.

ROAD STOP

24

How does energy power machines?

Muscle energy and mechanical energy supplied the energy for early tools and machines. Later, fossil fuels and electricity became the main sources of energy for powering machines.

1 KEEPING COOL
In ancient Egypt, wealthy people kept cool by relying on the muscle energy of their servants, who fanned them with huge palm leaves.

2 COOKING
In a microwave oven, a device called a magnetron turns electricity into a beam of short radio waves. A fan scatters the radio waves around the oven. The waves penetrate food and cause its atoms to vibrate rapidly. Friction among the moving atoms generates heat, which cooks the food.

3 TRANSPORT
In a car, an internal combustion engine burns fuel, usually gasoline, with air. The hot gases that form expand and push down pipe-shaped devices called pistons. The up-and-down movements of the pistons provide the power to make the car move.

4 LEISURE
Electricity powers many of the machines that entertain us, such as radios (invented in 1895), televisions (1920's), tape recorders (1940's), and personal computers (1970's).

WHICH COUNTRY USES THE MOST ENERGY IN THE WORLD? SIDE ROAD TO ROAD STOP 25

FOLLOW THE PATH TO THE **POINT OF INTEREST**.

 ### Visible Proof SPOT

See how friction creates heat. Generate friction by briskly rubbing together two blocks of wood for about half a minute. When you touch the rubbed surfaces of the blocks, they should feel hot.

POINT of Interest
Using energy that is millions of years old.

What are fossil fuels?

Fuels such as coal, petroleum, and natural gas are called fossil fuels. They are formed from the remains of plants and animals that lived millions of years ago. Burning fossil fuels releases the energy stored inside them.

COAL	PETROLEUM	NATURAL GAS
Coal is used mainly by power stations to produce electricity. The coal is burned to create heat, which changes water into steam that turns turbines. Coal is also used to heat buildings.	Most petroleum is used to make liquid fuels such as gasoline or diesel oil. These power cars, airplanes, and other vehicles. Some petroleum is burned as fuel in stoves and ovens.	Natural gas is found buried underground or beneath the seabed. Huge drills on a production platform sink wells to reach the gas deposits and pipe them to the surface.

Detour
BURNING FOSSIL FUELS RELEASES DANGEROUS GASES. LEAP TO **ROAD STOP 26**.

FOLLOW THE **HIGHWAY**. →

Detour
HOW CAN WE CONSERVE ENERGY IN THE HOME? LEAP TO **ROAD STOP 26**.

POINT of Interest
Using energy to say hello.

How does energy help us to communicate?

Every form of communication uses energy. The development of electricity revolutionized the way we send and receive messages.

Visible Proof SPOT

Today, electricity powers most machines found in a kitchen. Look at this picture. What would it look like if there were no electricity?

ANSWER: THE KITCHEN WOULD BE COMPLETELY BARE.

FOLLOW THE PATH TO THE **POINT OF INTEREST**.

SMILE!	TELEGRAPH	TELEPHONE
You use muscle energy when you smile at someone. Light energy transmits the image of your smile to the other person, who then uses the electric energy of neurons to convert the image into a signal that the brain understands as a smile.	The electric telegraph was the first form of communication to send messages using electricity. One of the earliest telegraphs, invented in 1837, had five needles that turned as electricity flowed through wires connected to them.	The American inventor Alexander Graham Bell (1847-1922) invented the telephone in 1876. A telephone works by converting the sound waves of a person's voice into an electric current, which the receiving telephone turns back into sound.

25

SIDE ROAD TO ROAD STOP 25

ROAD STOP 25

How much energy is used around the world?

Detour

COUNTRIES THAT USE VAST AMOUNTS OF ENERGY NEED TO LOOK AT WAYS OF CONSERVING THEIR ENERGY SUPPLIES. LEAP TO **ROAD STOP 26**.

The amount of energy a country uses depends on how its people live, what they can afford, and the kinds of machines and fuels to which they have access.

USING ENERGY

Many people in industrialized countries own cars, electrical appliances, and devices. Often people in developing countries cannot afford such things and few homes have gas or electricity. A person in an industrialized country uses about 15 times more energy than someone in a developing country.

Looking after the energy that powers our planet.

Select

YOUR SIDE ROAD

ENERGY IN THE **Future**

26 Why do we need to conserve energy? (HOW CAN WE CONSERVE IT?)

27 What do we mean by alternative energy?

28 How do we harness the power of the sun?

29 What is nuclear energy? (AND CAN IT BE DANGEROUS?)

30 Where might energy come from in the future?

ENERGY CONSUMPTION AROUND THE WORLD

IN INDUSTRIALIZED COUNTRIES, SUCH AS THE U.S. AND THE U.K., MOST ENERGY COMES FROM FOSSIL FUELS

- TRANSPORTATION
- INDUSTRY, FARMING
- HEATING, COOKING, LIGHTING
- FOOD

HOW ENERGY IS USED IN INDUSTRIALIZED COUNTRIES

IN DEVELOPING COUNTRIES, SUCH AS BRAZIL, PEOPLE DEPEND LARGELY ON MUSCLE ENERGY

- TRANSPORTATION
- INDUSTRY, FARMING
- HEATING, COOKING, LIGHTING
- FOOD

HOW ENERGY IS USED IN DEVELOPING COUNTRIES

TOTAL ENERGY USED BY AN INDIVIDUAL IN A YEAR

COUNTRY	ENERGY USED
India	0.6 million kJ
China	2.5 million kJ
Chile	4 million kJ
Australia	16.5 million kJ
United Kingdom	17.5 million kJ
U.S.	34 million kJ

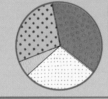

Strange! Some people generate energy from rotting plants and animal droppings!

ROAD STOP

26 Why do we need to conserve energy?

WHAT ARE THE WORLD'S MAIN SOURCES OF ENERGY?

- PETROLEUM = 40%
- COAL = 28%
- NATURAL GAS = 20%
- HYDROELECTRIC = 7%
- NUCLEAR = 5%

Extracting and burning fossil fuels can damage the environment and cause pollution. Fossil fuels are also not renewable, which means that one day they will run out.

FOLLOW THE PATH TO THE **POINT OF INTEREST** TO DISCOVER HOW TO CONSERVE ENERGY.

DANGER!

* Most factories and power stations generate energy by burning coal or oil. This produces waste gases that react with water droplets in the air. They form weak acids that fall to the earth as acid rain. Acid rain pollutes rivers and lakes and kills trees.
* Exhaust fumes from cars, trucks, and buses contain gases, such as carbon monoxide, that pollute the air.

DWINDLING RESERVES OF ENERGY

The world's increasing demand for energy is fast using up the supplies of fossil fuels. Even the discovery of new reserves cannot halt the decline. Experts estimate that petroleum may run out in about 2050, natural gas in about 2060, and coal in about 2200.

Visible Proof SPOT

Acid rain eats into buildings and statues. To show this, place a piece of chalk in a glass of vinegar. The vinegar is a weak acid and reacts chemically with the chalk. Carbon dioxide bubbles rise from the chalk as it slowly breaks apart. Eventually, the vinegar dissolves the chalk completely.

FOLLOW THE **HIGHWAY** TO DISCOVER HOW WE MAKE NUCLEAR ENERGY.

POINT of Interest
Everyone can play a part.

How can we conserve energy?

There are simple things that everyone can do to help save energy.

1 DON'T DISCARD – RECYCLE!

It takes a great deal of energy to manufacture new goods, yet people discard so much that could be recycled. Waste is usually dumped in large holes in the ground, called landfill sites (see right). It will stay in the ground for hundreds of years.

* Reuse or recycle paper, bottles, cans, and plastic containers.
* Mend broken objects instead of throwing them away.

2 SPENDING POWER!
* Buy environmentally friendly products. * Avoid heavily packaged goods.

3 TRAVEL RIGHT!
* Walk, bike, or use public transportation. * Share rides with friends.

4 DON'T BE WASTEFUL!
* Switch off the TV, radio, and light when you leave a room.
* Turn down the thermostat a notch or two.

HOW CAN WE HARNESS THE POWER OF TIDES TO PRODUCE ENERGY? **SIDE ROAD** TO ROAD STOP 27

CAN SUNLIGHT POWER A HOME? **SIDE ROAD** TO ROAD STOP 28

WHAT HAPPENS IN A NUCLEAR POWER STATION? **SIDE ROAD** TO ROAD STOP 29

WHAT IS A PERPETUAL MOTION MACHINE? **SIDE ROAD** TO ROAD STOP 30

ROAD STOP

27 What do we mean by alternative energy?

To meet the growing demand for energy in the future, scientists are exploring energy sources other than fossil fuels. This is known as alternative energy.

RENEWABLE ENERGY

Most alternative energy sources are renewable, which means that they will never run out. But more research is needed to make them efficient and able to produce the same amount of energy as fossil fuels.

TIDAL ENERGY

A tidal power station harnesses the energy in tides using a dam built across the mouth of a river. At high tide, water rushes through tunnels in the dam, and at low tide, the water flows back out again. The movement of the water turns turbines in the tunnels, which generate electricity.

The world's first tidal power station opened in 1966, on the River Rance in Brittany, France.

The Hoover Dam, on the Colorado River, has a hydroelectric power station that supplies most of the electricity for Arizona, Nevada, and southern California.

HYDROELECTRIC ENERGY

Hydroelectric power stations use the energy in falling water to generate electricity. When water stored in a reservoir is allowed to escape, it flows out with great force. As it falls, it turns huge turbines that are connected to generators.

→ FOLLOW THE **HIGHWAY** TO THE END OF YOUR JOURNEY.

ROAD STOP

28 How do we harness the power of the sun?

Every 40 minutes, the sun delivers as much energy to the earth's surface as the world's entire population uses in a year. People use this energy, known as solar energy, to heat homes and generate electricity.

Detour

SOLAR ENERGY IS SAFE AND CLEAN – IS NUCLEAR ENERGY? LEAP TO ROAD STOP 29.

GENERATING HEAT

Some buildings, especially in hot countries, have solar panels fitted to their roofs. The panels, also called flat-plate collectors, absorb heat from the sun and use it to heat water.

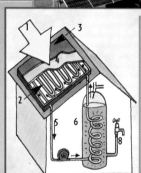

1 SUNLIGHT
2 SOLAR PANEL
3 CLEAR PLASTIC OR GLASS SHEET
4 METAL OR PLASTIC PLATE, WHICH ABSORBS SUNLIGHT AND CONVERTS IT INTO HEAT
5 WATER IN PIPES FIXED TO THE PLATE CARRIES HEAT AWAY
6 STORAGE TANK
7 WATER IN PIPE HEATS WATER IN TANK
8 WARM WATER FROM TANK IS PIPED TO TAPS

SIDE ROAD TO ROAD STOP 29
SIDE ROAD TO ROAD STOP 30

WIND ENERGY

Modern windmills, or wind turbines, use the kinetic energy of moving air to produce electricity. As the wind catches the turbine blades, they spin and drive a generator hidden in the turbine.

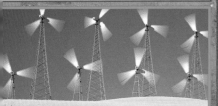

This "wind farm," in California, generates electricity.

BIOMASS ENERGY

Energy that comes from plants is called biomass energy. Burning wood supplies heat and light. Rotting vegetation, animal dung, and human sewage produce biogas, a mixture of methane and carbon dioxide. This gas can be burned and used for cooking and heating.

GEOTHERMAL ENERGY

The earth holds huge reserves of heat energy in hot underground rocks. This is called geothermal energy. Power companies can drill down to rocks that lie close to the surface and use the heat to generate electricity.

GEOTHERMAL POWER STATION

GEYSERS

The pressure from hot underground rocks can naturally force boiling water out of the ground. It explodes as hot springs or geysers. In Iceland the superheated water is used to heat buildings and supply hot water.

1. HOT ROCKS, WITH TEMPERATURES OF UP TO 1,832° F (1,000° C)
2. COLD WATER IS PUMPED DOWN A BOREHOLE
3. THE WATER BOILS AS IT MEETS THE ROCKS
4. STEAM IS FORCED UP ANOTHER BOREHOLE
5. THE STEAM DRIVES TURBINES THAT POWER A GENERATOR

 Visible Proof SPOT

In a geyser, hot rocks heat the water so that it boils. Bubbles of steam rise to the surface and force the water out. To show this, stand a funnel in a pot of water. Put a rubber tube under the edge of the funnel and blow hard. Air bubbles force water out of the top of the funnel.

Detour

IS THERE ENERGY IN SPACE? LEAP TO ROAD STOP 30.

FOLLOW THE **HIGHWAY**.

On a hot day, sunlight passes through the windows of a car and warms up the air inside. Many buildings heat air in a similar way. They have large windows that face the sun. The stone or brick walls and floors absorb the heat during the day and release it at night.

HOW IS SOLAR ENERGY CONVERTED INTO ELECTRICITY?

Solar cells, also called photovoltaic cells, turn solar energy into electric energy. Photovoltaic cells are often found in satellites in space and solar-powered calculators.

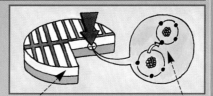

A solar cell is usually made from two layers of silicon. One layer has extra electrons, while the other has a shortage of electrons.

When a light shines on the cell, it forces free electrons from one layer to move to the other. This flow produces an electric current.

Satellites are powered by hundreds of solar cells. They are often grouped together in large flat panels.

 Visible Proof SPOT

Dark colors absorb more light than light colors, which is why solar panels are black. Put one cube of butter onto a piece of white cardboard and another onto a piece of black cardboard. Place both pieces in the sun. The butter on the black piece melts quicker because the black color absorbs more of the sun's rays and turns them into heat.

IS NUCLEAR ENERGY RENEWABLE? **SIDE ROAD** TO ROAD STOP 29

WHAT FORMS OF ENERGY MIGHT WE USE IN THE FUTURE? **SIDE ROAD** TO ROAD STOP 30

ROAD STOP

29 What is nuclear energy?

Nuclear energy is the powerful energy that is produced by changes in the nucleus of an atom. Nuclear energy can be used to generate electric energy.

The first commercial nuclear power station opened in 1956, at Calder Hall, England.

WHAT HAPPENS INSIDE A NUCLEAR POWER STATION?

- **REACTOR** A tanklike structure, with thick, steel walls that absorb radiation.
- **COOLING TOWER** Steam or water vapor escapes through these towers.
- **TURBINES** Water is turned into steam, which drives turbines, producing electricity.
- **PYLONS** These carry away electricity to be distributed.
- **CORE** This contains the fuel, in which the nuclear reactions take place.
- **HEAT EXCHANGER** This transports the heat energy, created by nuclear reactions in the core, to a supply of water.

FOLLOW THE PATH TO THE **POINT OF INTEREST**.

NUCLEAR FISSION

Nuclear power stations produce energy by nuclear fission.

1. A slow-moving neutron hits the nucleus of a uranium atom.
2. The nucleus splits into two smaller nuclei and releases more neutrons.
3. These neutrons split more atoms, and so on, in a chain reaction.

Visible Proof SPOT

Fire a marble at a group of marbles packed together. Most of the marbles stay put, but some roll away. In nuclear fission, neutrons release other neutrons from atoms in a similar way.

FOLLOW THE **HIGHWAY**.

POINT of Interest
Sometimes, things can go badly wrong.

Can nuclear energy be dangerous?

A nuclear power station produces radioactive waste, which is highly dangerous. Much of it is buried in pools of water at the reactor sites.

CATASTROPHE ... In 1986, an explosion at the nuclear power station in Chernobyl, Ukraine, released huge amounts of radioactive material into the atmosphere. Up to 8,000 people died as a result of the disaster, while thousands more developed cancers and other illnesses.

ROAD STOP

30 Where might energy come from in the future?

In the future, scientists may be able to produce energy from sources other than water, wind, or hot rocks.

SPACE ENERGY

One day, scientists may build huge power stations in space. They would consist of satellites, several miles across, covered in solar cells. These would collect the light energy radiating from the sun and beam it down to the earth.

Energy makes things happen.

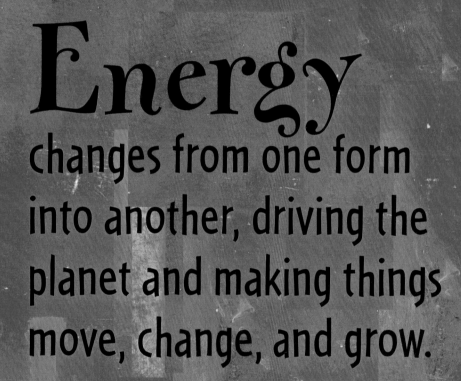

Energy changes from one form into another, driving the planet and making things move, change, and grow.

NUCLEAR FUSION
The sun uses nuclear fusion to produce a vast amount of energy. Scientists have not yet worked out how to produce efficient fusion on earth. If they succeed, they will have discovered a reliable source of energy that produces no radioactive waste.

PERPETUAL MOTION
For centuries, people have tried to invent a machine that, once started, would never stop. It would run without any outside source of energy. No one has ever succeeded in building such a machine, and almost all scientists believe that no one ever will.

Think!
The energy that helped to create the universe still exists today, although it has been changed into different forms many, many times.

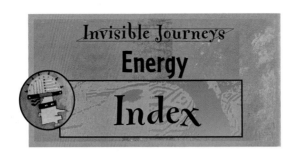

Invisible Journeys
Energy
Index

acid rain 27
acupressure 12
acupuncture 12
alternative energy 28, 29
amber 19
Anaxagoras 13
Aristotle 13
atom 7, 8, 13, 14, 15, 18, 24, 30

bacteria 11
battery 7, 18, 19
Bell, Alexander Graham 25
big bang 8
big crunch 8
biomass energy 29
boat 10, 23
brain 20

calorie 5, 11
car 24, 25, 26, 29
carbohydrate 11
carbon dioxide 9, 29
carnivore 11
central heating 24
chemical energy 7, 10, 11, 17, 19
Chernobyl 30
chi 12
chlorophyll 9
circuit diagram 18, 19
coal 25, 27
combustion 24
communication 25
conservation of energy 17
consumer 10, 11

de Coriolis, Gaspard 14
decomposer 11
Democritus 13

Einstein, Albert 15, 17
electric circuit 18
electric current 15, 18, 19, 20, 21, 25, 29
electric eel 21
electric ray 21
electric telegraph 25
electricity 6, 7, 15, 17, 18, 19, 20, 21, 24, 25, 28, 29, 30
electrocardiogram 21
electrocardiograph 21
electrode 19
electrolyte 19
electron 6, 15, 18, 19, 29

Faraday, Michael 20
farming 10
fat 11
fire 23, 24
firework 17
fission 7, 30
food 5, 7, 9, 10, 11, 17, 24
food chain 11
force 5, 13, 16
fossil fuel 24, 25, 26, 27, 28
Frankenstein 21
friction 17, 24, 25
fuel 7, 23, 24, 25, 26, 27, 28, 30
fungi 11
fusion 7, 31

galaxy 8
Galilei, Galileo 13
gasoline 24, 25
generator 20, 21, 28, 29
geothermal energy 29
geyser 29
glucose 9
gravity 16

heartbeat 21
heat 6, 7, 8, 11, 14, 17, 18, 23, 24, 25, 27, 28, 29, 30
helium 7, 8
herbivore 11
Hoover Dam 28
hydroelectric energy 20, 27, 28
hydrogen 7, 8
hypocaust 24

impulse 20
internal energy 24
ion 18, 19

joule 5
Joule, James 5

kinetic energy 6, 13, 14, 16, 17, 20, 23, 29

Leucippus 13
light 6, 7, 9, 10, 11, 14, 15, 17, 18, 23, 24, 25, 27, 29, 30
lightning 17

machinery 23, 24
magnet 20
mass 14, 15
matter 8, 17
microwave oven 24
muscle energy 5, 6, 10, 11, 16, 17, 24, 25, 26

natural gas 25, 27
nerve 20
neuron 20, 25
neutron 7, 30
Newton, Isaac 14
nuclear energy 17, 27, 30
nucleus 7, 8, 18, 30

oxygen 9, 10, 23

Penzias, Arno 8
perpetual motion 31
petroleum 25, 27
photoelectric effect 15
photon 15
photosynthesis 9
Planck, Max 14, 15
plant 7, 9, 11, 29
plow 10
pollution 27
potential energy 7, 10, 13, 16, 17
power 5
power station 20, 21, 25, 27, 28, 30
producer 10, 11
protein 11
proton 7
pyramid 10

quantum 14, 15
quantum theory 14

radiation 6, 8, 14, 30
radiator 24
radio wave 6, 8, 24
reflexology 12

sail 10, 23
satellite 29, 30
soil 9, 11
solar cell 29
solar energy 15, 28, 29
solar panel 28
solar system 8
sound 6, 7, 17, 25
static electricity 19
steam 20, 25, 29, 30
substation 21
sun 6, 7, 8, 9, 10, 11, 17, 28, 29, 30, 31

telephone 25
temperature 24
Thales 19
Thompson, Benjamin 14
thunder 17
tide 28
transformation of energy 16, 17
transformer 21
transmission line 21
turbine 20, 21, 25, 28, 29, 30

universe 8, 13, 17
uranium 30

vacuum 7
velocity 14, 15
volt 19
Volta, Alessandro 19

water energy 5, 20, 23, 28, 29
water wheel 23
wave 5
Wilson, Robert 8
wind energy 5, 10, 23, 29
windmill 23, 29
work 5

X ray 6